ORIENT EXPRESS

ORIENT EXPRESS

Poems by
GRETE TARTLER

Translated by
FLEUR ADCOCK

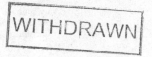

Oxford   New York   Auckland
OXFORD UNIVERSITY PRESS
1989

Oxford University Press, Walton Street, Oxford OX2 6DP
Oxford   New York   Toronto
Delhi   Bombay   Calcutta   Madras   Karachi
Petaling Jaya   Singapore   Hong Kong   Tokyo
Nairobi   Dar es Salaam   Cape Town
Melbourne   Auckland
and associated companies in
Berlin   Ibadan

Oxford is a trade mark of Oxford University Press

First published as an
Oxford University Press paperback 1989

British Library Cataloguing in Publication Data
Tartler, Grete
Orient express: poems.—(Oxford poets)
I. Title   II. Adcock, Fleur, 1934–
859'.134
ISBN 0–19–282699–9

Library of Congress Cataloging in Publication Data
Tartler, Grete.
Orient express: poems / by Grete Tartler; translated by Fleur Adcock.
p.   cm.
Translated from the Romanian.
1. Tartler, Grete   Translations, English.   I. Adcock, Fleur.   II. Title.
859'.134—dc20   PC840.3.A76A22 1989   89–3379
ISBN 0–19–282699–9

Typeset by Wyvern Typesetting Ltd.
Printed in Great Britain by
J. W. Arrowsmith Ltd., Bristol

ACKNOWLEDGEMENTS

Some of these translations have appeared (a few of them in earlier versions) in: *The Honest Ulsterman*, *Illuminations*, *Numbers*, *Poetry Review*, the *Times Literary Supplement*, and *Writing Women*.

Oxford University Press acknowledges the financial assistance of the Arts Council of Great Britain towards the publication of this volume of translations.

FOREWORD

I first met Grete Tartler in 1984, in Bucharest; I was there on a British Council cultural exchange visit, and she introduced herself to me at a gathering in the Writers' Union and presented me with a book of her poems. At that time I had no knowledge of Romanian, but I proceeded to learn it (mostly from books), and by the time I visited Romania again in 1987 I was able to show several of the women poets I had met, including Grete, sample translations of some of their poems.

I came away with more books of poetry, and of these I found that Grete's work seemed to lend itself most naturally to translation into English—or, at least, to translation by me. This may reflect my limited understanding of Romanian idioms and poetic practice rather than the relative qualities of the poets concerned, but I think it is true to say that her poetry, while by no means lacking in subtlety, is less obscure than that of some of her contemporaries. It was certainly a pleasure to discover it, poem by poem, through the process of bringing it into English.

Grete Tartler was born in Bucharest in 1945; her family is from Transylvania and, as her name implies, she has German ancestry. She studied music at the Ciprian Porumbescu Conservatorium and later obtained a degree in English and Arabic from Bucharest University. For fifteen years she taught the viola in a secondary school, but she recently left to become an editor with *Neue Literatur*, a review of German literature in Romania. She has published seven collections of poetry (one of which won the Poetry Prize of the Writers' Union) as well as a number of translations from classical Arabic literature. She is married to the novelist Stelian Tabaraş and has a daughter.

My selection of her poems is taken from her two most recent collections, *Substituiri* (1983) and *Achene Zburătoare* (1986), both published by Cartea Românească.

I am deeply grateful to Andrea Deletant for reading through my original versions of these poems, pouncing on two or three howlers, and offering useful advice at several other points. She is not, of course, responsible for such infelicities as may remain.

FLEUR ADCOCK,
London, October 1988

CONTENTS

ORIENT EXPRESS

PHILOLOGY

Imagine you suddenly rise up into the air
and from there you can see everything—
the network of silver railway tracks from which
not a single event is ever derailed.
In autumn we're going to learn Sanskrit together.
Although we speak the dialect of grass
and of the hot streets, of burning rubbish;
although my body is as luminous
as glow-worms when they mate (an initial letter
so brilliant that you took it for a fire
and were about to stamp it out);
although we've patiently looked up in dictionaries
the three thousand words for 'misfortune',
there's one hope left to us: Sanskrit.
Perhaps this is the way to understand
what poisoned fish say in the waters
of a river on whose banks we once held hands;
perhaps this is the way to understand
the parched fields, the parched roof of the mouth,
the ruins of the Second World War,
the bunkers of the Third,
speeches on 'the battle is life and life is the battle'.
  How afraid I was, as a child,
  of the open mouth of a bell!
  Today I live between its jaws:
  the heavy bronze tongue beats my eardrums,
  beats me on the mouth, on my spread fingers . . .
  a smooth trickle on my chin, a secret letter:
all we need to do now is decipher it.

THE FLYING CARPET

In summer on the shore of the lake
surrounded by bluish-white tower-blocks
some people are coming outside
to sunbathe, and others to wash carpets.

And I remember the little girl
who jumped out of a window, clutching
a doormat she believed was a flying one
(which it may have been, since they caught her in time).

But you're clutching excitedly
at scorched grass, lying flat on the ground,
and you can see through screwed-up eyelids
a red ball bouncing on a wall.

What's more, in the bazaar you've bought
a watch that makes time stand still.

AERIAL PHOTOGRAPHY

Almost everything you hear
could sound different. The ice on the lake
could crack with a sound like a brass band,
the crocuses could open with a drum-roll,
the worms gnawing the truth underground
could sound like thunder.
The voice of a man
returning from the sea with his trawls empty
could foam with rage.
But look at the price of transforming rage to patience!
    In the park an old man is kneeling
    to photograph the bells of snowdrops.
    Not long ago, in a fighter plane,
    he was photographing the cities of Europe.

BITTER WATERS

She gets out of the dusty red car
and stops by the well:
her mind is an invisible shore
under the play of spiderwebs.
She drinks the water and it seems bitter—
it's the vessel of her body that's bitter,
tortured through so many lives
into so many strange shapes:
it still keeps the taste of death.
She drinks the water and it seems bitter.
She counts on her fingers
the things she has renounced:
the city,
       friends,
              uninhibited laughter,
all for a tiny, useless miracle.
I watch her as she leans over
combing the grass with thin fingers,
and I should like to be the one
who's still waiting for miracles.
                  Secretly
I throw down the gold ring.
She finds it, smiles, is certainly happy;
but she doesn't straighten up:
she combs the grass with her fingers;
she goes on looking;
she goes on looking.

SIGHIŞOARA 1982

If you had the insight
as the river has, flowing between
yellow and green houses with chained doors,
breathing out two breaths—the scent
of lime trees and that of damp—
listening to the organ-music of tiles:
    piano (the ones with moss on them)
    forte (the redder ones).
If you had the impulse
as the cannon do, booming
    over ANDREAS RATH, IRONMONGER,
    over the high school, over the volley-ball field,
    over the jasmine and gnomes in the garden.
If you would bring supplies
from the other side of the river!

To be clouded and indifferent as silt!
You have lived many lives here in vain.
The water carves a scar
on the face of the town;
even you have set the gold ring
from your finger under your eyelids now:
a snake painted on the chemist's shop wall—
you read there 'Water Level, 1970'.
But you, so indifferent, don't bring
supplies, although you've been crossing
to and fro across the river for so long.

AT THE RACECOURSE

Leaning on the iron railing, the young wife
sees lorries at the abattoir unloading
files of cattle. A tongue of red, to the east,
uncovers the jagged teeth of the city.
Ah, now that they've finally moved
to a new flat, if only they could have
next to their block a clean, green racecourse!
If the abattoir could be moved; if she could sleep!
She could invite her friends on to the balcony
to watch the races; they'd clap, they'd enjoy it.
And she'd no longer have to wedge
a pillow around her head, or turn
her cassette-player up to its maximum volume
to reduce the day's bellowing
to a tolerable rattle in the throat.

BAROMETER

'You don't bathe twice in the same sea'
I tell myself, in the amniotic foam
beside the jetty, under a bone-coloured sky.

Instead of a liver, a mirror;
under my eyelids, a ship with no anchor;
God struggles in Heaven only to break my heart.

You look through me and no longer notice me.
A cold sweat covers the planes in the hangars;
small ghosts wriggle through my body, through the crystals.

And nothing is visible in me but my breakdown;
nothing is visible but how I try to show
the pressure, until the last moment.

THE EMPTY SHELL

On the shore, among broken glass and bottle-tops,
an empty snail shell.
I leave litter too—my initials
on you: your ear's a G,
over which my finger writes a T.
A log-book: after 57 days
  and a journey of 3270 miles
  in the papyrus bark Ra II,
  we have seen no trace of man—
  only, everywhere, his rubbish . . .
After the journey
with no supplies but your unbowed back,
after the long pilgrimage
in search of the Mirror,
only this empty snail shell:
a leftover. If you haven't the courage now
to throw it into the flames,
how will you dare to shed your shroud
when the day comes?

BURNT GIBLETS OF GAME

The boys put kebabs on the fire,
laugh, drink bilberry wine, taste the mushrooms
they picked in the evening. Is love to be
this, burnt giblets of game?
I stalked you like a true hunter
for many centuries. But today
I sail through everything, like the clouds.
In the mountains—well,
here you can retain your ideals.
As evidence there's the forest inspector,
who has spent two years on singing lessons
and once knew Galaction.
As evidence there's the chalet attendant
who won't pose except in a suit.
We take photographs, make kebabs, drink beer;
then walk through the cemetery in the valley
among the work of photographers
more amateur than ourselves.

ETERNAL LIFE

Some of the pensioners are at table,
happily gobbling brains with egg.
Others, on the station benches,
are reading novels and slanging passers-by.
A greenish vapour rises from the pages
between hearts like Yo-Yos, leaping and falling.
'Carry on, read some more!' says an old woman.
But her old husband stands up suddenly,
as if drawn by a magnet.
He flaps his wings above her—she can see him—
but he's beyond her reach now.

DON'T WORRY

'Don't worry!' he tells me, in the brass market.
'Don't keep life chained up so tightly!
I had a chain once that was off a crane—
I kept a mastiff on it—and it lasted five years.
I had an Armenian brass one; it broke after three.
The chain around the monument
needs to be replaced less often—
but the monument isn't always fretting at it!'

'Don't worry,' I reply, 'the chain of love gets stretched,
but in sleep it knits itself back into place.'

SHE TALKED SOMETIMES

She talked sometimes about the Shah of Persia—
how she had sung one evening in his presence,
how the diamonds and emeralds had glittered in the dark.

I didn't know her mezzo-soprano voice,
but in autumn I'd seen feet on slender heels
pattering up the stairs with a bird's gait.

After a while she stopped appearing: day and night
from the locked attic there's been heard only the rattle
of a typewriter, driving the pigeons away.

Can she be learning to type? Is there no more demand
for singing lessons? Is she writing and burning her memoirs?
From under the door creeps a thread of smoke, of grey hair.

THE SPOKESMAN

You're playing the viola in the greenhouse on the enclosed
  balcony;
your wrist, a waterfall, moves to and fro
above the pallid stalks, above the snapdragons
('If you see them grinning, don't think they're smiling at
  you'),
above the soil babbling about the loves of the Others.
Here you are far from the world, and yet
life suffocates you, as if someone had smashed
a bottle of perfume in a bus—
you'll have to get off at the next stop,
or else it will make you sick.

IN THE LIFT

You get into this musical instrument.
The air, heavy with so much breath,
climbs together with you to the top floor.
They told you 'There's not even a crack.'
But all the same, from somewhere a breeze comes in,
a dark wing, vibrating.
The top button is missing. The lift sways,
pendulum-fashion,
as if a leaf shifted from left to right,
from one paving-stone to another;
as if a shiver passed
from your flesh to that of your twin.
Forget that you're going home; burn your wings
as soon as possible!
Perhaps as early as tomorrow a storm
will lift the lid off the box;
and then, don't stay any longer
stuck between narrow walls!

BUTTERFLY AND CANDLE

Closing-time, when hairdressers cut each other's hair
and in the markets the sweepers pile vegetation
into yellowed grave-mounds—
passer-by, weep for the death of the goddess!

Only the jars of honey still glow, dimly.

But my face is visible in you
as the stone is visible inside a cherry,
as ash is visible at the tip of a flame.
'She's discovered something', people say to each other,
'but what can it have been?'
In the flame, only she knows.

THE WILD STRAWBERRY

Soon the spider will be tapping messages on the walls,
and on the balcony of the flat you'll be watching over
the wild strawberry you brought back from a holiday—
the pentagram of leaves over its forehead.

The girl's feet are walking across the asphalt
in harmonic sympathy with the blackened weeds.
A raven appears to her left; the wind's on her right;
her body leans to the left, her mind to the right.

Soon, open-mouthed, you'll be breathing the magnolia—
its white clavicle, its copper seal.
Soon you'll be standing below it with your head back;
you'll see people walking past, leaning right back.

LIGHTS UNDERFOOT

With what satisfaction, today,
the street gulps down every mouthful.
A thousand puddles you've never crossed;
a thousand districts where you've never walked.
What's the use of putting on
your white shoes, lights under your feet?
Your heart shrinks. Only the shoelaces
are still worth using now.

And what's the use of your poems? To leave
a mark in the dust when you lift them from the shelf?
Like a group of apricot trees
saved from the debris, beside a blue verandah.

Soon this old district
will have been completely demolished.

HOMUNCULUS

The gardener inserts pumpkins and cucumbers
into narrow-necked bottles to grow;
oblong pears and even bunches of grapes
swell inside their containers
like homunculi in test-tubes—
then everyone wonders how such opulent fruit
could fit into any bottle.
In the same way my fists, like two waterlilies
floating on a marsh, have grown
from infancy inside a crystal covering.
Now they shine and even open out
at night, in the moonlight,
like two glasses whose contents
are gradually becoming clear.
It's hard to say which you see first,
the bottle or what's in it,
the blood, or what it is that makes you bleed.

LANDSCAPE WITH PUPPETS

Snow makes the mountains visible—underlining
the words. Meaning disappears
in the uncertainty between the lines.
The tourists queue up, motionless,
at the funicular railway station,
each with a single hope, knowing
they're pulled along on cables, like puppets.
In light which recalls Daniel Turcea's
determination to write, after
he had learned what his end would be,
make snowballs to build a wall!
Bury in the foundations
an empty rum bottle
with a message in it: what you believe
about the truth, about people.
Write the message, but without
underlining the words.
Then, on top, heap up as many snowballs
as you can. Frozen snowballs.

THE WONDERS OF THE WORLD

Tell me about the wonders of the world: a glow-worm
which feeds on earth
but never eats its fill
for fear that all the earth will be eaten up.
And think of those who sell themselves to death:
the mercenaries.
(But what a difference between 'ray' and 'raid':
for a moment you're frightened.)
And once in summer there was a swirl of schoolboys
in red blazers on the stone steps
with a scent of lime-trees over the scent of damp,
and weeds, jasmine, garden gnomes,
peasant women with wide straw hats making a ring of shade
like a space for prayer.
Or think of the cricket who sheds his skin
(the way you found him in the doorway)
before he begins to sing.
If you're hungry for wonders, stuff yourself with these
while they've not yet disappeared under a scab of grass,
while among them words too can flash like lightning.
Perhaps you will remember
these moments of gluttony
in the future, when you're asked 'Who—
your parents, your child, your great love, who—
taught you that poetry was where to look
for your ultimate salvation?'

WINGED SEEDS

The 'Papercraft' studio on the ground floor has closed.
At first we were glad: cheap tinfoil stars,
party decorations, jolly colours—
in spring the goods were displayed in the open air,
propped up around the trunk of the plane tree
with its winged seeds.
Then the studio expanded; they bought a small shop
on the other side of the street,
then another, and another—a row, a whole quarter
of prosperous shopping-space.
Gaudy plywood flowers tempt you in.
In the end, the plane tree in the centre
was felled. Why should we trouble ourselves
with questions about these winged seeds—
where they come from, where they are going?

THE SIMPLIFICATION OF VOCABULARY

It's snowing for the simplification of vocabulary.
The sparrow is happy that it's caught a spider,
the schoolchildren that they've caught a tram.
Tariq ibn Ziyad has been rejoicing for a thousand years
because he conquered Gibraltar.
And I hold you happily by the hand.
Eternal vanity, to believe,
between the sides of this bag of fluffy rot,
that I'm in control of an illusion—
that I can break out of the globe of water
I live in, where it's beginning to snow
(if you turn it over and shake it, that is);
that we can say 'dreamers' instead of 'poets',
'Marx' and 'Freud' for 'this century's gods',
'accident' instead of 'unhappiness'.
How sensually we shall relish, under our camouflage of
    snow,
the simplification of vocabulary!
(But only if you shake the globe, of course.)

THE WATER-LEVELS OF THE DANUBE

On the radio they're broadcasting the water-levels of the
   Danube.
I stoop and take up water in my hands—
as much as I can use, save, waste:
the caesium-limit (9, 132, 631, 770 parts) of a second.
The events of the day almost merge into the background,
like a flock of pelicans: a white circling,
an over-dull sequel to an over-long epic.
The fact that I have never worn a watch
on my wrist, not since the Conservatorium years
when I was learning the viola for hours on end,
is that really a sign of not adapting? 'We'll meet tomorrow.'
But it's long ago that tomorrow was the time
when we'd meet under the monastery gateway
and you'd take off your armour, and I'd have a cure
for wounds, a bottle of poison.
We have no watches; we don't even know our age.
And now I take my direction from
the flowering of a waterlily, the rustle of reeds, the scents of
   the jungle.
You welcome the rain—'You're here, my lady!'—
with a ritual; you spread out paper reservoirs
to be filled with the shower that irrigates poems.
And it is only a handful of water
(as much as I can use, save, waste for you)
in an indifferent river.

WITH FOUR WINGS

While you're walking in full sunshine
I'm beside you in a shower of rain;
you're breathing the chill of a world you can't see.
Don't you feel my clothes brushing against you,
soaked by the clouds of my hidden world?
Aren't you drinking lightning from between my teeth?
We're on the same road but quite separate from each other.
The road's marked out; you won't change the signs now.
It's our daily road, past the market-stalls—
I see two birds tied together:
although they have four wings now, they can't fly;
but they gaze at each other at close range.
   'You have the eyes of an owl.'
Yes, an owl which doesn't live in a cage.

FLOWERS OF THE FIELD

At the blood-test laboratory
people come out one by one
each carrying a test-tube of blood
like a lighted candle.
(Could you carry a poem in your hand like that
without spilling a drop?)
Prickly smells, faces anonymous
as camomile, but plucked out and chosen.
The colours show up as clearly
as in a painting.
(Remember the well-known question:
'Which would you save from a fire, a famous picture
or a man who hadn't long to live?')
Reflections from the test-tubes make faces blush red,
illuminate rough hands.
This is surely a picture that must be saved:
a procession of poppies
through the long corridors,
advancing, invading—making no mistake
about how long they've got.

MAKING A POEM

A day in May,
and I rush into a field of poppies
as if pleading for initiation.
An eddy of fumes:
the plants make me dizzy; my arms
float up almost at once, like pages of writing,
a notebook taking off in flight.
You laugh and say to me:
'Stay for three days with your eyes closed,
without eating or drinking; and then
cook huge plates of delicious food—
if you're still willing to leave them alone
for the sake of these wild fumes, then,
yes, you'll be allowed to try.'

TROPAEUM TRAIANI

Bones, silver coins
swimming as in an aquarium—
and the perforated metope used for ages
by peasants as a stone for a well-head.
Here they draw water
straight from history.
Don't lean over! You yourself,
dying of thirst,
asked me for a well by the roadside.
But now, throwing yourself into it,
you scream and blame me.

ARTIFICIAL BLOOD

Artificial blood has been invented.
You, who have been both stone and flower,
will be what you can't imagine.
You know nothing about yourself,
you are wax in a honeycomb;
when you are a candle you will scarcely know.
But what can you know? The questions, wool on the spindle,
enclose us; they will use us up.
The wick is threaded through the candle—
or are we being spun alive
by the blind spinning-woman?
The questions are the same, although
artificial blood has been invented, although . . .
You, who have been both stone and flower,
will be
what you can't imagine.

ALCAHEST

Autumn is coming; I'm learning to make
the Alcahest which dissolves matter.
I transform my liver, heart and spleen
into planets;
I listen to their rolling, then they give birth
to a golden mask—no, a heap of them.
The words which died have left in their place
these graven images.
I thread them up, and gaze for a while at the thread—
ah, the thread, which everyone knows
doesn't become a jewel itself
simply because it's threaded through pearls.
My realist friends would snap it with their teeth.

THE VIOLIN-MAKER

With his wire spectacles and shiny bald head
in a room full of canaries,
he was the most famous instrument-maker,
and my parents had asked him
to make me a violin.
I remember: I was a schoolgirl; I climbed
the narrow dovecot staircase
next to a basket on a rope, intended
for hauling up mail or newspapers.
Then I became a student, and still no violin—
there was still some detail, something to finish.
Today my daughter needs it,
but the violin-maker frowns:
'I won't be hurried
just because people happen to inherit
an obsession with playing an instrument.'

THE SEALED BOOK

Waiting for you I play 'Tunnel of time'.
Somewhere at one end
your ancestors embrace each other.
'Hurry,' I call to them, 'I too have learnt
the secrets of music, the art of concentration
(but for any kind of teaching
many other arts are needed—
it's no use trying to go on milking
a goat that's knocked the milk-pail over.')
Then, at the other end,
your ashes envelop me.
Suddenly I discover the secret
of the sealed book
which I'd been carrying in my body.
Then you come home. You're right on the threshold.
The leaves of the book inside me
thrash around in a great storm.

THE COANDA EFFECT

I cross the bridge after the train:
vertigo like a tongue of flame
appears on the crown of my head—
the thoughts of others, of the passengers
who are asking me for a miracle.
Each of them wants a concrete miracle,
each wants something different—
at least groves of pomegranates by the track,
at least golden bridges through the pomegranates.
I could prove to them that what they'll get
are tricks, little fairground deceptions:
anything else, I'd shout to them, is a miracle.
But the train has gone. The dusk is pouring
quicksilver into my blood.
I pull my collar over my ears,
listen, and stagger across the bridge.

SUFI LEGEND

A curious story:
of how a Sufi arrived in the land of the mad
who were frightened of a melon-plant,
believing it was a green dragon.
The Sufi cut down the melon
but was in his turn cut down by the madmen,
who were even more frightened
of a dragon-killer.
I would have won their confidence
and agreed that it was a question of a monster.
In fact that's what I've done—
then I gradually convinced them that it's a melon,
and taught them how to cultivate it.
Now the plant has covered my window;
a yellow fruit is growing on it, swelling—
it's making my room dark.
Although I'm not a sage, I shall have to cut down
both the stalk and the yellow skull
which grins at me in the sky.
If the plant and the monster were flattened
I should have much more light
to discover for this poem
an optimistic ending.

THOMAS TARTLER

I defended this town for years and years:
I ran the secondary school, I taught
the students to believe in *unitas populi*;
I was present when the guild-workers
declared war on the Austro-Hungarian emperor.
They put us in chains, they whipped us into cellars,
and then they led us to the scaffold.
The soldiers shouted at me 'Hurry along there!'
'And what about you?' I said to them.
'Would you hurry along and run to your death?'

It's certainly a splendid town,
a tourist attraction, full of historic remains.
The guide leads me through the square
where the scaffold used to stand,
and asks me please to hurry along.

THORNS

I'm warming my hands
at the fire in the market
in a tin bucket; the peasants throw
paper, vegetable-scraps, twigs into it.
This is how the Romans ended up.
Sometimes the dead person
was still alive, but the mourners
convinced him he was dead.
'I'm alive!'—I struggle and push the onlookers
out of the way—'It's a mistake!
And the fire,' I shout at them, 'put it out!'
But if we've been brought roses
what are we to do with the thorns?
I know a bird (on the way to extinction)
which keeps its prey spiked on thorns.

THE BELL FOUNDRY

On Sunday we go for a walk
to Plumbuita, past the bell foundry.
The chestnut leaves are drumming a tattoo.
There's a buzz of dandelions, dragonflies and frogs
making our ears tingle. Heads of corn,
a crowd of them, clash together slowly.
One day I tried to lift the giant bell
overturned in front of the dog-kennels,
and out of spite because I couldn't shift it
I gave it a kick.
The rumble frightened me, of course,
but why should I be the only one
to put my hands over my ears?
The swift ropes of silver have caught up with us.
Two great mirrors are crushing us between them.

CAPRICCIO

With the boards left over
from repairs to the house
I built myself a boat on the balcony.
Waves of winged seeds, blown on the wind,
lure us on, beckoning to us.
I push aside the lock of hair
on my forehead, the imagined truth,
my prejudices.
A magnet pulls the nails out
from the boat, the yellow dress from the line.
My mind is being inlaid around you
like mother-of-pearl inlaid on a shell.

MEDIEVAL NOVEL

For years on end I studied the secret
of gothic courtyards shining after rain.
A bucket of water:
when I shake it
the old woman at the window is my daughter;
she looks at me with big eyes, without a smile,
without a word, without anything that's 'her'.

Do you remember the Saxon porcelain-maker
who betrayed the secrets of the glaze to his daughter
out of love?
And she, out of love, betrayed them to a young man,
then flung herself into the burning kiln . . .

For years on end I learned to kill dragons;
but I didn't meet any dragons: that skill
was never required of me. All the same,
I never betrayed the secret.
Now I pull out weeds,
great leaves full of sap, which quiver
when I fling myself down among them.
We both have dark circles under our eyes.

In the oak tree next door
a nightingale has begun to sing.

LACE MUSEUM, BRUSSELS

Lace from Flanders, Binche, Chantilly, Malines,
point Rosaline, point d'Angleterre:
I jot down the forgotten names, with the thought
that I'll write something about the Middle Ages;
but while I'm scribbling I feel a surge of anger
that bodies are born ready perforated,
ready riddled
by gunshot, by radiation—
weeping circles
trying first just to get used to life
and then to get used to lack of life.
God, how stupid lace is,
toiled at for years and years, hung
in front of windows, plugging up
dustbins, cannon-mouths—
a mass of circles through which you pass
like a tiger through hoops of flame!
What an illusion, that you could fill up the circle
before it fills itself up.

AFRICAN VILLAGE

A smell of Nile-water, goats and smoke.
The beehive huts tempt me;
I find the narrow entrance.
Pigeons flock over the place I vanish into.
And in the centre
only a brass jar
and an almost newborn baby girl,
her eyelids black with kohl.
The women paint my eyelids,
put the brass jar in my arms,
sing and clap their hands around me.
Chains glitter, flashing harshly;
the jar is my skull of brass.
I tip it up beside the single tree.
I tell them I have come to water
this unique tree; I fill the jar
again, over and over—for a lifetime—
and yet my footsteps
leave almost no trace in the sand.

THE COLLECTOR

Never mind about wind, dust, death-rays: it's time for
the Sappho butterfly ('Why is that name familiar?'
wonders the entomologist's wife.)
The collector heads for the fields, armed
with nets and encyclopaedias:
he's happy chasing after the blue wings—
he forgets about the black ones;
and our friend the conductor is a great fisherman,
and another friend has an amazing stamp-collection.
They've escaped from between the jaws
of ideas; they're enjoying themselves on earth—
'Let us get on with our lives too, in our own way.'
The professor of harmony comes in; we talk about
Challenger, about Eliade's death:
'He scattered his ashes on the wind like a Hindu.'
'How important the youth culture is.'
Outside, the pupils we've polished laugh; there's a pause;
marble dust comes in through the windows—
no, it's fluff from the poplar trees;
or it's butterflies,
or perhaps flakes of ash
looking for the Collector.

OPUS MULIERUM

Old courtyards with tubs of laundry:
'Go to the washerwoman and do your own washing'
I whisper to you, and the wild apricot trees
all turn suddenly white, the sky pales,
the world is soused in a drenching buzz.
There's a smell of bluebags and a sulphurous bubbling.
You'd hardly believe it—so much steam rises
that only dirt is left in the copper.
The wild apricots petrify into coral.
It's so easy—easy in a woman's way—
to wash your soul, to rejoice in the spring wind
shaking the scales on its dragon-tail
so that you're looking at soap-bubbles
it blows for you between your fingers.
Two children pass by, holding on a string
a balloon transparent as a bubble.
For a moment we are crouched inside it.

HALLEY 1986

I'm well aware it's only a passing flame;
but I'll bury containers of contemporary evidence
(for possible extraterrestrials)
and rent myself an island in the eastern seas
where I'll cultivate the explosion of the senses
or the retreat into the self (those two paths
that are free to all, libido and religion,
have phonetic elements in common.)
It won't be the end of the world
because we, 'the young poets', are still
in the future; but through fear of the comet
there are no more suicides, even of those
whose only way it is to get into the papers.
So the days pass. Is it coming?—Or not?
Until then, I'll choose a jug from the heap
because the potter's voice was engraved on it,
as on a record, while it turned on the wheel.
Until then I'll listen to the angry murmur under the glaze
     and I myself, turning,
     may be a container
     in invisible hands.

SAHEL

Each year the desert
                    (beginning with *d* for 'destiny')
advances fifteen kilometres
                    (beginning with *k* for 'karma'),
dries up the streams
                    (beginning with *s* for 'spirit'),
dries up more and more words.
The dictionary is gradually being starved—
in mid-leap, essences
halt for a moment on the brink of the abyss,
then their bones bleach on the cracked earth.
The poet watches
the clean skulls of the words.
The words, still alive and ravenous,
watch the poet.

ORIENT EXPRESS

Nearly asleep, I'm reading the Desert Fathers.
There are towns, turquoise plains;
at stations I hear announcements in unknown languages.
A man in my compartment was in the war;
he used to play the trumpet.
The woman next to me is crocheting (knots
between good and bad, between truth and falsehood).
It's as if I'm conducting the rhythmic pulse of the train,
the chorus of those who are staying awake
for fear of the dawn.
Once, on holiday in the mountains,
I heard this train go by,
the one I'm on today;
someone told a story about the snake that sucked from a cow:
the men found it asleep among the rocks
and struck it with an axe;
milk flowed from it as if from a cask.
Now over the hills, over the acid waters,
over the Greenpeace ships, over the explosions,
the small publicity, the smog, the dried-up springs,
milk flows in waves.
I feel it taking the form of hills,
the form of the brain:
dawn flows into it without filling it;
dawn leaves it without emptying it.

OXFORD POETS

Fleur Adcock

James Berry

Edward Kamau Brathwaite

Joseph Brodsky

Michael Donaghy

D.J. Enright

Roy Fisher

David Gascoyne

David Harsent

Anthony Hecht

Zbigniew Herbert

Thomas Kinsella

Brad Leithauser

Herbert Lomas

Derek Mahon

Medbh McGuckian

James Merrill

John Montague

Peter Porter

Craig Raine

Tom Rawling

Christopher Reid

Stephen Romer

Carole Satyamurti

Peter Scupham

Penelope Shuttle

Louis Simpson

Anne Stevenson

George Szirtes

Grete Tartler

Anthony Thwaite

Charles Tomlinson

Chris Wallace-Crabbe

Hugo Williams

also

Basil Bunting

W.H. Davies

Keith Douglas

Ivor Gurney

Edward Thomas